Legal Boost: Big Profits Through an IT Transformation

Tom Lambotte

Braeburn Publishing

Twinsburg, Ohio

Expert Technology Strategist for Mac-Based Law Firms:

Tom Lambotte

Legal Boost: Big Profits Through an IT Transformation

GlobalMac IT
7851 Bavaria Road, Suite B
Twinsburg OH 44087
United States
440-252-4600
www.GlobalMacIT.com/boost

Helping Mac-Based Law Firms Eliminate Technology Headaches Finally and Forever

All rights reserved. No part of this publication may be reproduced or transmitted in any form by any means, electronic or mechanical, including photography, recording or information retrieval system, without written permission from the author.

Printed in the USA.

Copyright © 2016 GlobalMac2, Inc.
Legal Boost: Big Profits Through an I.T Transformation/Tom Lambotte.
- 1st ed.

Dedication

To my wife Christine – it's a privilege
to share my life and love with you.

To my children Eliana, Atreyu and Malia, who will inherit
this world and make it a much better place, your growth
provides a constant source of joy and pride.

*If you haven't found it yet, keep looking.
Don't settle. As with all matters of the heart,
you'll know when you find it. And like any great
relationship, it just gets better and
better as the years roll on.*

— STEVE JOBS

What can you expect from "Legal Boost: Big Profits Through an IT Transformation"?

"Tom Lambotte is one of the top IT consultants working with lawyers who use Macs in their law office. As both a practice lawyer and the organizer of MILOfest, the only conference for Mac-using attorneys, I get to see the entire landscape of providers. What Tom does, and how he does it, is the cream of the crop. This book is full of practical advice to help lawyers in the area of IT management. Even if you think you know it all, there's something here for you. You should read this book, and then you should talk to Tom about having his company, GlobalMac IT, help you with your IT needs."

- **Victor J. Medina**
  Managing Attorney-Medina Law Group
  Founding Organizer - MILOfest

"It's tempting, as a Mac-using attorney, to try to be your firm's technical support department as well. In this book, Tom Lambotte makes the case for why doing that is *exactly* the *wrong* thing to do. As veteran provider of information technology support services to lawyers using the Macintosh platform, he knows what he's talking about. And you should listen."

- **Mark C. Metzger**
  Attorney

"What can I say, Tom just "Get's it." He not only is a master of all things Mac, he intimately understands how a law firm works, what an attorney's IT needs are and why. As a contributor to our monthly marketing brief for entrepreneurial attorneys, Tom consistently provides real world advice to our members and we believe his firm is an essential choice for a Mac-based law firm."

- **Richard James**
  President, Automated Business Results
  Law Firm Marketing Consultant

CONTENTS

Foreword 1
Introduction 3
The Hidden Costs of Wearing the IT Hat 7
The Cost of Slow 17
Features Are Worthless 23
007 At Your Service 27
The Price of Ignorance Exceeds That of Education 31
Process Makes Perfect Part I: External 37
Process Makes Perfect Part II: Internal 45
What Options Does a Mac-Based Law Firm Have For IT Support? 55
What Aare Mac-Based Law Firms Saying About GlobalMac IT? 63
Incredible Free Special Offer… 72

FOREWORD

Some people work in their business, while others work on their business. Do you know the difference? You should, because one can quickly lead to burnout, while the other can pave the path to success. Learning the difference sooner rather than later can lead to a more successful and satisfying career.

In practicing law for over two decades, I've been fortunate to have had the opportunity to meet many successful businessmen from all across the country in a variety of industries. Because I was a financial management major in college, I have always been interested in what makes successful people tick. One common trait that I have observed in these industry leaders is their understanding of the important role that processes play in developing a successful business.

Process is defined as a series of actions or steps taken in order to achieve a particular end. Note that it doesn't mean "the way you do something." Instead, the key concept to grasp is that the actions taken are geared toward a specific goal. Working for its own sake is little more than spinning one's wheels, which leads to a lot of action but no progress.

Instead, thought leaders take the concept of process one step further and continually refine their goals, so that they are forced to constantly improve while simultaneously being kept from the dangerous trap of resting on their laurels. When Steve Jobs decided to change the way we all communicate by developing the iPhone, he wasn't satisfied with the unprecedented success of the first model. Instead, he pushed his developers to constantly refine and improve it and he continued to do so until his last days.

Tom Lambotte understands the importance of developing and refining processes, and his relentless pursuit to make his business the best it can be will motivate you to want to do the same. I have known Tom for several years, and I have had the pleasure of giving presentations with him and watching him present. I can say without hesitation that when it comes to legal technology and the ways to use it to improve your law practice, Tom knows his stuff.

Lawyers, as a group, are not the most tech-savvy people and we often tend to be thrifty (i.e., penny wise and pound foolish). Across the board, we tend to be at the tail end of most technology trends and are often the last to incorporate them into our practices. To make matters worse, in many firms, particularly smaller ones, lawyers tend to attempt to manage their own technology needs because we mistakenly think that it's less expensive for us to do so.

What often gets lost in the mix is that we went to law school to learn how to practice law, not how to manage the technology needs for a business. Even worse, the time that we spend trying to keep abreast of technology changes and implementing them into our practices (in the best case scenario) or trying to figure out how to resolve problems when they arise (more typically) is all time that we're not practicing law.

Most attorneys still bill by the hour, so they undoubtedly understand that time is money. However, whether it's pride, stubbornness, cheapness, or a combination of these, too many lawyers forget the time = money equation when it comes to running the business side of their practices. I hate to say that, but I say it because I believe it's the truth. Why else would smart attorneys spend time trying (for free) to solve problems for which they are not trained instead of hiring an expert to do so and spending those hours practicing law? When you look at it that way, it just doesn't make sense, does it?

After having several long conversations with Tom at a legal technology conference, my firm decided to let Tom's company, GlobalMac IT, handle our IT needs. Not only do they resolve issues when they arise, but they also handle routine matters such as adding new employees and monitoring our systems proactively to keep them secure and to address any problems before they arise.

I can easily say that going with GlobalMac IT was one of the best decisions we have made, as we freed up (at a minimum) several hours per month dealing with IT issues, and we are now able spend those hours serving our clients. This not only results in significantly less stress for us, but also in happier clients and increased revenue as a result of more billable hours and more time spent working on our practice.

www.GlobalMacIT.com/boost   440-252-4600

So, as you open this book, consider whether you are satisfied to simply work in your business or if would you be better off learning how to develop and refine processes to be the best that you can be? If you choose the former, I wish you the best of luck, because you will probably need it. If you choose the latter, turn the page begin learning from Tom's techniques and approaches so that you will know how to reach your full potential.

Ben Stevens
Family Law Attorney
Publisher, The Mac Lawyer legal technology blog
Spartanburg, SC

INTRODUCTION

In the legal world, technology is too commonly viewed as an expense which should be minimized. It is seen as a necessary evil, one of those "costs of doing business" that we just have to endure. In my world, where we only work with Mac-using law firms, the overwhelming majority try to do everything they can to spend as little as they have to on technology. I've read hundreds of posts about keeping a 6-year old Mac running by swapping out the hard drive. That doesn't sound like a good way for an attorney to spend his time.

The problem is that there is a lot of 'penny-wise and pound foolish' behaviors going on. I want to highlight a few numbers everyone should be aware of, as they will be crucial for you to keep in mind as I attempt a mental transformation for you in this book in how you view technology in your practice. In 2015, according to Forbes, Legal Services is the 2nd most profitable industry in the United States, with an average of 17.8% profitability. This industry also has the 4th highest cost of payroll, coming in at 41.8%. Ok, so what? Well, what I take away from the first number, based on almost a decade of supporting Mac-based law firms, is that you can do a horrible job running a practice and still make money. This is quite a forgiving industry: you can have no systems and work with antiquated technology and still make a profit. A double-edged sword. At the same time, payroll is your biggest expense.

What you need to realize is that technology plays a huge role in your firm's profitability. How? Think of how much time everyone on staff spends working on their computers. Are there old, slow systems? Is there a good practice management solution in place? Stable email, shared calendars? What's the experience when someone needs to work remotely? Who jumps in when an issue pops up?

Is it the managing partner, who puts everything on hold and chats on a user forum to figure out how to get his internet back online when his whole firm is down? An Office Manager who should be implementing a marketing campaign which would create leads, bring in new clients and boost revenue, but puts that on hold, to fix a printer issue? Or your staff, who resorts to 'just dealing with it' and finds some workaround because they a) don't want to interrupt you since they know you have much more important things to do or b) have zero confidence in the IT guy?

In this book, I am going to provide some paradigm shifts that will transform the way you look at the role of technology in your practice and offer you suggestions on how to change technology from being an expense you need to keep to a minimum, into an investment you make in your firm, that will substantially increase your profits, reduce bottlenecks in your firm that are costing you tens of thousands of dollars and boost everyone's productivity. If your payroll is the biggest cost, won't increasing the work output make a significant increase in your profits? Any added output while keeping costs constant goes straight to your bottom line. And technology can have a major impact here.

Having supported Mac-based law firms from coast to coast for nearly a decade, I have seen it all. I started as your standard, hourly IT guy and grown and evolved into the premier option for Mac-based law firms. Through our evolution we have completely pivoted how we approach and deliver IT services; going from a reactionary method of support (when doing hourly) to a proactive method, in ways most IT providers only dream of. The stability of our solutions and complete end-to-end, white-glove solutions for the Mac-based clients we serve has allowed us to create raving fans and make significant impact upon their firm profits, all while developing long-lasting friendships.

In this book, you will learn 6 of our core concepts that will allow you pivot the role IT plays in your firm, allowing you to invest more time on growing the firm and less time putting out fires. I promise that if you follow and openly consider the ideas shared in this book, that you will walk away with some actionable ideas you can implement which will make a positive impact on your firm's bottom line.

Don't be the person who shuts down because "I use Macs because they just work, so this doesn't apply to me." The ideas presented here are NOT about the Macs. They are about the systems and overall strategies in how you should approach the role of technology in your firm. Be the kind of person that takes these to heart, takes action and benefits from the results.

The tips you're about to read have been proven to create positive, long-lasting results for the firms in which we have implemented them. All you have to do is block off the time and commit to reading through these 6 core concepts. Grab a highlighter and a pen, mark up the book, and take away what strikes home. Each chapter will introduce a new way of looking at technology and its role in driving your firm's profitability and overall productivity. Take control of your technology and dive in!

CHAPTER ONE

THE HIDDEN COSTS OF WEARING THE IT HAT

"I am providing this post to advise my fellow family law attorneys of a huge mistake I made for the first two decades of my legal career – handling most of my firm's IT needs myself.

For instance, when one of our computers had a problem (and when we used PCs there were many), I thought the best course of action was to try to solve it myself instead of having a professional do so – as it would 'save money.'

All of that changed late last year when we (finally) realized that the best course of action was outsource certain functions in order to allow us to spend more time doing what we do best – handle family law cases. We started using GlobalMac IT for our computer and network needs, and we are beyond pleased with the results."

>Ben Stevens, prominent Blogger at
>www.themaclawyer.com and
>Partner at The Stevens Firm Family Law Center in Spartanburg, South Carolina

The following white paper addresses many of these misconceptions that could be costing your firm more money than you ever imagined:

Are you wearing the IT hat? Know someone who is? If so, I'd like you to consider this article an intervention, an attempt to compel you or that someone to get help. I define "one who wears the IT hat" as the person who is the primary go-to person for IT issues in their firm. This individual deals with the bulk of IT issues and pride themselves on the cost savings in doing it themselves. They often have an hourly IT person to call in dire needs.

The hidden problem is that the costs of taking care of IT yourself for your firm, far outweigh the perceived benefits. In most cases, this is, in fact, costing your firms tens of thousands of dollars per year. Most people calculate the perceived savings by simply looking at their P&L and seeing close to nothing on the line for IT services. This overly simplistic way overlooks the soft costs which quickly eclipse the cost of paying for proper IT services.

There are four primary concepts that need to be understood in order to help overcome and break down the limiting belief that doing IT yourself saves money. The first is an analysis of time spent by the one wearing the IT hat. The second attempts to quantify the hidden issues lurking in the shadows and their overall impact on employee productivity. The third relates to what I call the hobbyist principle. And the final one is the far-reaching impact of time spent (or not spent) on the highest ROI activities for the firm.

TIME ANALYSIS

The time analysis takes two parts into consideration. The first is simply the time spent per month dealing with IT related tasks. In most cases, the one wearing the IT hat is a senior partner or founder of the firm, since they set things up from the start and are the only one that understands how everything is setup. The time he spends dealing with IT primarily consists of running software updates (Adobe, Microsoft Office, Java, Flash, Apple/Windows software updates and security updates, etc.) and basic, daily troubleshooting. Based on many years of supporting law firms, this is an average of 1-2 hours per computer, per month. For a firm with a staff of 5, that is 5-10 hours per month.

The second part of the time analysis is the impact of interruptions. Because everyone in the firm is dependent on this person when there is an issue, they are forced to interrupt that person to get their issue resolved. Some are quick, easy ones (5-15 minutes), some take more time 'Googling' around (15-45 minutes), and others take hours before eventually being given up on, at which point the IT person is reached out to (and is hopefully able to help in a timely manner). All these little interruptions add up. Let's whip out our calculators here to properly comprehend the impact. Research shows that when interrupted, it takes an average of 23 minutes to get back to the previous task. Let's be ultra-conservative and use

15 minutes and one interruption per day, which adds up to 300 minutes, or 5 hours per month.

Adding up the two parts of the time analysis above shows us that the person handling the IT in a small firm with a staff of 5 is actually spending about 10-15 hours per month on IT. With a conservative rate of $300 per hour, that is a loss in billable wages of $3,000-4,500. One more point about time, viewing it from a different angle, is how it touches on work/life balance. You see, in most cases I have come across, the person wearing the IT hat cannot justify doing the updates and system management during prime daytime hours, so they end up doing it in the evenings or on the weekend, either way, they are taking away irreplaceable personal and family time.

"WE'LL JUST DEAL WITH IT"

Without fail, every firm we have supported in which someone had been wearing the IT hat, there has always been a laundry list of problems that eventually surface. These quite often had been masking issues that could develop into bigger, more impactful problems which could easily be avoided with proper maintenance. Not once have I come across a firm where people sit around twiddling their thumbs waiting for something to do. The support staff and other attorneys are well aware that the person running the firm and, in this example also wearing the IT hat, has both important and urgent items that need be addressed that take precedence over dealing with IT issues. This develops an "I'll just deal with it" mentality; if they can find a workaround to the IT issues, they will seldom report it and just deal with it. Because of this, when we take over IT support, we are very proactive in coaching and teaching everyone in the firm to tell us about every issue, big or small. As this coaching process is repeated on our end, people always step forward with things that have bothered them for years. Small, medium and big issues that have just been dealt with.

Let's whip out those calculators again and calculate what the cost is of the impact on your staff's productivity. Payroll is, for most firms, the biggest cost by far. If people are wasting 10 minutes a day due to bugs in the setup, inefficiencies with the server, calendar, email, printers, etc.; things they have just found workarounds with, that adds up to 200 minutes a month per person. With 5 people, that is 1,000 minutes or 17 hours per month. What is your firms' average fully burdened cost for a staff member? At a very conservative cost of $50 per hour, that is $850 a month. Over the year, that is $10,200.

THE HOBBYIST

Jack-of-all-trades, master of none. As cliché as this is, it is so true. When an attorney is taking care of IT themselves, they will always be a hobbyist and hence, never develop mastery. In most cases, how do things work out when your clients try to represent themselves and practice 'Google" law? I'll go out on a limb and assume these often end up being major 'cleanup jobs,' where it could have taken only $1/10^{th}$ of the effort if you had taken on the case from the beginning. Why is that? Because you have developed mastery in your domain over the years, which allows you to assimilate all the details and think of the majority of possible angles. A 'greener' associate does not have the insight and experience as a seasoned attorney. Developing this mastery takes years and thousands of hours of focus. And one of the biggest benefits to your clients which you bring and they will never be able to is all the things they don't know they don't know. The same applies to IT. You will never be able to develop mastery in IT when it is one of the many things on your laundry list of responsibilities.

The cost of being a hobbyist with your IT can be massive. The things you don't know you don't know will be long. For example, it could end up costing you your license, by not being able to convince a board of ethics that you took the proper preventative measures in securing your client's data. It can cause hours or days of downtime for your firm with an issue that could have easily been prevented with proactive maintenance. It can cost you all of your client files and data because you setup a backup yourself that was 'automatic' 9 months ago, but did not realize the drive became disconnected 5 months prior.

Hobbies are fun, but I would not choose to be a hobbyist with something that has a huge financial impact on my business and personal life. Most people use a tax accountant because they are up to date with the current tax law, know what questions to ask and how to maximize their deductions. TurboTax cannot provide me with these insights. I do not want a tax hobbyist taking care of my taxes, nor attempt doing them on my own. I want someone who has developed a mastery in tax law, and that I can count on when I need guidance with a tax related question or event. I surround myself with experts and have built a team of superstar advisors because I understand the value that can be added using this approach. If you want a hobby, pick up cooking, fishing or golf, but leave the things that have a big impact on your firms to experts.

ROI IN IT

What activities within your law firm generate the highest ROI? Driving the vision, focus and direction of the company? Working on the most profitable cases and clients? Or is it interrupting time spent on the above to run a software update or fix an email problem someone in your firm is experiencing? This last and final cost, the opportunity cost, is rarely calculated. The cost of taking care of IT yourself is far, far greater than just the fact that you are spending 10-15 hours per month on IT. If you invested these newly freed up 10-15 hours per month on your highest ROI generating activities, what impact would that have on the bottom line of your practice?

In closing, I hope this intervention has been successful in helping you question previously limiting beliefs you or someone you know may be holding onto, by doing the IT for their firm themselves. Having an expert IT company with developed best practices for your firm can shift the role of IT from just something that has to be dealt with, to something that adds value and profitability to your firm. Through a holistic approach, the right IT firm can provide proactive support and implement solutions that can impact your bottom line and free up you and your staff's time. Through a two-prong approach, they can first seek out, then address the root cause of the primary bottlenecks in your operations, help hidden issues surface to the top and address security concerns and implement better procedures. Then they can look at your firm as a whole and work on implementing solutions that increase everyone's productivity.

CHAPTER TWO

THE COST OF SLOW

A very strange thing occurred in October 2014. A prospective client signed up for our services, wrote us a big check, then disappeared. When he came to us, he had been reading our materials for some time and was ready to move forward. He definitely had a need and was keenly aware that his IT situation was out of whack. Being a Mac guy himself, he had been trying to make the move from PCs to Macs for some time. He had accepted that he wouldn't be able to do it himself, saw the value in our services and we had a signed agreement overnight. We charged his card and started our in-depth onboarding process. Then…he disappeared. Completely. Not a peep. We couldn't get a callback and were unable to connect via calls or email. We eventually gave up and were left scratching our heads as to what happened.

Fast-forward three months later, and out of the blue, I received an email from him saying he was ready to get started. We were happy to have his business, but I was more concerned about what had happened in the first place. I had to clear the air and be confident he would prove to be a good, responsive client for us to work with. Our service depends on having a quality two-way relationship with our clients.

He explained to me that after we spoke his father had convinced him that "if it ain't broke, don't fix it," and that since the computers were still running, why upgrade them? These were old PCs running Windows XP and Windows 2003 Server. Well, after that decision, two systems went down, one of them actually went down twice, and one never came back up at all. This left their users unable to work for multiple days at a time, and one of his employees was literally left without a computer to work on. So he was now more ready to move forward than ever before, especially since they were tax attorneys heading into the busiest season of the year. We picked things up and since then have switched his office to Macs and everyone is now thrilled.

In terms of technology, the old adage, "if it ain't broke don't fix it" is an immensely flawed belief system. There exists a perception that the longer you keep computers, the better 'bang for your buck' you are getting. Technically that is true. However, the unrealized cost of this is massive. Today, I'm going to dive into this topic through 2 primary arguments. First, I am going to challenge the "perception of savings," helping you see the real cost of slowness. Second, we're going to dive into the value of your time.

Based on a 2007 survey, the U.S. Census Bureau report shows that the legal services industry has the 3rd highest payroll costs, coming in at 45.03%. If you can increase your firm's productivity, you can increase your profitability. That sounds like more money in your pocket. Ok, so how do we impact productivity using technology?

In its most basic method, I always start by first looking at bottlenecks that can be removed and then determine ways to increase productivity through training and having the proper tools and solutions in place. Without getting too geeky (this is not the place), today's fastest consumer CPU (processor) is over 3x as fast as the fastest consumer CPU that was available 5 years ago. In the meantime, we also moved from DDR2 to DDR3 memory, and SSDs (solid state drive) are slowly becoming the norm over HDDs (hard disk drive). The experience of a new high-end PC today is definitely a whole lot different than five years ago.

That time spent waiting for Internet Explorer to startup, or the time it takes for a computer to startup and log in? You're paying for that, and it is a lot more than you realize. If a modern computer saves someone just 10 minutes a day, that comes out to 2,400 minutes per year, or 40 hours. On average, what is the fully burdened cost of one week of payroll? Let's take $40 per hour as a very conservative example, which adds up to $1,600 per year, per computer. This isn't even taking anything else into account, such as having an easier User Interface to work with, as you get with a Mac, additional training that can be provided to your staff, or rolling out solutions that make it easier for your staff to achieve the things they do dozens of time every day.

You see, slow is expensive and in most cases our computers are the bottlenecks in our workflow when we are waiting for them to finish, waiting for programs to open, for large PDFs to be OCR'd after they are scanned (you do scan all your files, right?), or waiting for a file to be saved. When working on a slow machine, most people don't notice it, since it's all they know, and they have become accustomed to it. Your role is to make sure your staff has the best tools available to them so they can be as productive as possible. Their job is not to tell you their computers are old and slow and need to be replaced.

SPEND MONEY WHERE YOU SPEND YOUR TIME

Once you've saved money, where should you spend it in order to maximize the usefulness of your money spent—or even your happiness? To answer that, just look at what you spend your day doing, proportionally, and allocate money accordingly. Let's call this the comfort principle.

Simply calculate how much of your day is spent using a certain item. Let's say we get our recommended 8 hours of sleep per night, leaving us with 16 hours. If we spend 8 hours of our day on or at our computers, we've spent 50% of our days on our computers. Let's get some actual numbers instead of abstract numbers, as this tends to be easier to grasp. In being (very) conservative I am going to use a 40-hour work week, and we'll say an attorney spends 30 hours of that time on their computers. Over 52 weeks, this comes out to 1,560 hours. Our company cycles all of our clients' hardware on a 3-year cycle, as we have found this to be the optimal schedule. Going with that, you will spend 4,680 hours on your computer over 3 years. Wouldn't it make sense to invest a few extra dollars there? Can you see how working on an old machine, simply because it still runs, is not a good idea?

Seeing how much time you actually spend working on your computer, wouldn't it be worth it to spend a little more upfront to make sure you have a great computer? If your computer takes 10 seconds to open an app, and you can shave it down to 2 by upgrading to a newer computer, that's a worthwhile purchase when you factor in frustration and time saved. If your computer locks up frequently because you don't have enough RAM (random access memory) or if it's just too slow, it's in your own interest to upgrade or get a new computer. If you and your staff can get through your day with as little aggravation, frustration and discomfort as possible, everyone will be much more relaxed, which benefits you and everyone around you. And preventing stress is much better than having to spend money later on to alleviate stress. A new client recently told us they had lost an Office Manager the year prior due to issues with her slow computer that were never addressed.

The sad truth is that most people don't "get" this concept. Time and time again, I have found myself talking to some clients with older systems, quite often the managing partner, trying to convince them to upgrade to a newer one. When the day comes, and they finally upgrade, I ALWAYS get a raving "thank you" and "I wish I had done this sooner." The same goes for laptop users finally getting a big monitor to work on, but that's a topic for another month.

Look at the hardware that you and your staff are working on. Do they have the right tools to complete their work in an efficient manner or are their systems creating bottlenecks in their workflow? Is your firm's hardware over 3 years old? If so, while it may feel good to not be writing a check for the computer hardware, you are, in fact, writing the check, it's just going to another line item, your payroll.

I challenge you to question the way you look at the cost of IT. Is it an expense you dread putting any money towards? Or is it a worthwhile investment that will increase your staff's overall productivity? Remember, the legal services industry has the 3rd highest payroll cost at 45.03%! Every boost in productivity for your staff raises your revenue and your overall profitability.

CHAPTER THREE

FEATURES ARE WORTHLESS

PCLAW. TABS3. PRACTICEMASTER. AMICUS ATTORNEY. NEEDLES. TIME MATTERS. TIMESLIPS.

Most attorneys will have heard of at least a few of these solutions as they have been the main contenders for decades. The problem is that they have turned into massive and immobile pieces of bloatware. What is bloatware? From Wikipedia: "In long-lived software, perceived bloat can occur from the software servicing a large, diverse marketplace with many differing requirements. Most end users will feel they only need some limited subset of the available functions and will regard the others as unnecessary bloat, even if people with different requirements do use them."

The problem with adding on every feature request for 5, 10, 15 years or more is that features, when you are not able to easily access them, become worthless. The software becomes so intimidating and non-inviting to the user that it simply goes unused.

You see, if you and your staff are not both comfortable and confident with a solution, it is not going to be used.

I know many firms who had been using TABS3, PracticeMaster or PCLaw. However, they continued to track time on an excel sheet, which they would then send to their Office Manager once a week to manually enter. I have also heard from clients, who are making the switch from PC to Mac, that although they own one of these archaic pieces of bloatware, they rarely use them, if at all.

Another very common problem I see for these massive companies, especially those that have been around forever, is that they are very slow in moving to the cloud. If they do decide to move, they simply copy what they currently have add a new pretty coat of paint and then offer it to their clientele. The problem here is that none of the weight is shed. This causes the migration to the cloud to be both slow and unexciting.

THE CLOUD

The new cloud-based management solutions have made gigantic strides in reinventing the software to what it should have been in the first place. They are lean, simple to use, easy to deploy and have great educational content to help the new users ease into it.

The older big name management solutions mentioned above often cost thousands of dollars to get up and running, thousands of dollars to maintain, and require expensive servers on-site, which in turn create expensive IT bills to properly manage and keep up to date. The new cloud-based solutions are turn-key, without massive initial fees to get up and running. They can easily scale and contract with the needs of your firm, making the process very easy to budget, since you will also know exactly what a new user will cost to add on. They also offer may ways to learn the software, so that when you bring on a new attorney, they can learn through self-paced methods such as pre-recorded videos on specific topics live webinars, and easy to search and use online support portals. In comparison, most of the older software will have a 300-page PDF manual, which no one ever cracks into.

One of the biggest features is that 'modern', reinvented solutions also can integrate with many of the solutions your firm currently uses. Services like File sharing, like ShareFile, Dropbox for Business, Box, Quickbooks Online, Xero (accounting software), payment gateways, and more can seamlessly integrate and talk with these case management solutions. These integrations will make working easier for your staff and will remove many of the tedious steps that are currently being manually performed by someone in your office.

If you feel your firm is currently using one of these 'bloatware' case management systems, I highly recommend taking a look at Clio and Rocket Matter. These are the two most popular 'modern' case management systems that our firm implements for our clients. The easier the interface of your case management solution is, the easier it will be for your staff to begin using it. The more they use it, the more productive they will be. And, whammy! You've just increased the profitability of your firm.

CHAPTER FOUR

007 AT YOUR SERVICE

The senior partner of a firm had just called our emergency line and told us his laptop and iPhone were stolen just minutes ago from a small café in Paris, France (not to be confused with Paris, TX). I couldn't help slightly grinning, knowing all of his client data was both completely secure and backed up, thanks to the tools and best practices we had in place, and having complete certainty that he would be fully operational and back to work within a few minutes. Rare moments like these, where we get to use the full capability of our tools, makes everyone on our team feel a bit like 007.

We had him enroll his wife's iPhone into our Mobile Device Management solution and tapped into the magical powers of our IT solutions with the click of a few buttons on our end. POW! Within less than 5 minutes, our client had full access to his email, contacts, calendars, case management system and firm files. The work he had been working on minutes before the event were all backed up, and he picked up right where he left off. Best of all, this was all accomplished with very little effort on our end and done in a calm and collected manner.

This last part is key, as we know what it's like to receive stressed out, unconfident support on the other end of the line. This allowed our client to remain calm and comfortable through the process, as we put our tools to use from the other side of the globe. He went back to work and enjoyed the rest of his workstation; our clients' trip was shaken a bit, but not stirred.

Disasters happen. Unplanned events. "Acts of God." This is why we carry at least half a dozen types of insurance. We do so to have peace of mind and certainty in times of chaos. You see, when something happens, your reaction should be calm and collected, "Well, that sucks, but I have a plan for this in place." For example, if you lost your phone or dropped your laptop, what would happen? Well, there's the financial hit of the loss of hardware, but the data you've most recently worked on is, in most cases, irreplaceable. In addition, the risk of a data breach is not to be taken lightly. When you've got the right IT Best Practices, support and solutions in place, your response should be something like this: "Oh, I lost the hardware, but the 'stuff' is secure, backed up AND I have solutions in place to get back up and running with minimal effort."

In this scenario, GlobalMac IT was able to save the day. We calmly and confidently explained to our client that:

- All of his firm and client data on his laptop and iPhone was completely secure, and there was zero risk of a data breach.

- Everything he had been working on all the way up to the fateful restroom break was backed up to our cloud file system and backup solution.
- He would be back up and running in a matter of minutes.

The role of IT in the majority of small to medium law firms is often not taken seriously enough. For example, we see far too many firms using one of Amazon's most highly rated and inexpensive, residential router or an Apple Extreme Base Station, to protect their firm's network. We commonly see people doing their own IT or receiving support from their neighbor's son living at home. You cannot expect someone whom you pay hourly and only in dire situations to implement pro-active solutions. Everyone has come across a legal case in which had they called you 6 months earlier, their case would have been much easier to take on and win. It's the exact same with IT.

I challenge you to question the role of IT in your firm and how it is currently being approached. Is it a reactive model? Ignore-it-and-hope-everything-is-fine-model? "If it ain't broke, don't fix it" model? – (with 7-year-old XP machines that are both a security risk and a massive loss leader in productivity in your firm). If you had been the one traveling in Paris and had your laptop and iPhone stolen with all your firm data on it, how would you have reacted and how quickly would you have been back up and running? James Bond did not make or guess which tools he would need on his missions, Q always provided them ahead of time.

CHAPTER FIVE

THE PRICE OF IGNORANCE EXCEEDS THAT OF EDUCATION

A runner for your firm is taking the scenic route every time he goes to file papers downtown. Let's assume it's not intentional, it's the just the way he's always gone. It takes him 45 minutes round-trip each time while a more direct path would take only 30 minutes. You are essentially paying him 33% more than needed, to complete the same task, which he does, daily. Doesn't seem like a big deal, but over the course of a year, you are paying him an extra 60 hours because he completes the task in an inefficient manner. Simply showing him a map (or providing a GPS) and a better path would allow him to complete 60 hours of other work while keeping your payroll costs the same.

According to Financial Information company Sageworks, in 2014, Legal Services was the 2nd most profitable industry, with a 17.8% Net Profit Margin. That's great but is a double-edged sword. You see, this means you can run a sloppy law firm and still make money. Another point to take into consideration is that the legal industry has the 4th highest payroll costs, coming in at 45.03% of revenues. Because this is such a huge cost, anything you can do in your practice to increase output while keeping your payroll costs fixed, will increase your bottom line profits.

Many businesses and law firms will address the problem of being busy by throwing people at the problem. That is a very expensive approach and helps keep those payroll costs up. However, if you have 10 people in your firm, and you can make them 10% more productive, you have just increased your work output to be what you'd create with 11 people on staff, but did so without increasing your payroll costs. So how can you do this? Focus on processes instead of bodies. What I'd like to discuss today the value of training your staff with the technology they use daily.

One thing I love about only supporting Mac-based law firms is that we spend very little time doing reactive support, where we go into fire-fighting mode. Our tools and solutions we implement reduce these to be extremely rare events. This frees up my staff and the staff of our clients to focus on more important things. Instead of putting out fires, they can focus on skill development. Instead of calling our support line to remove a virus, they call us for IT training and questions. We also have a lot of proactive and ongoing training for all of our clients so that they can continually get better at the things they do

Abraham Lincoln said, "Give me six hours to chop down a tree and I will spend the first four sharpening the axe." One of our 4 Core Values at GlobalMac IT is to Sharpen the Axe. We try to get everyone on our team to spend 2 hours per week on skill development. The more they do this, the better they become, the higher their output is, the better communications and experiences our clients have with them and so on. The added value of skill development is almost impossible to calculate since its impact touches so many things throughout your business.

By investing in your people, you can increase their output while keeping your payroll costs the same. So where do people spend a lot of time? There is nothing in the office that your staff spends more time on than their computer. That is why I discussed in a previous article the "Cost of Slow" and how critical it is that your staff has access to fast, modern computers to assist them in completing their work. Now, let's say you took my previous article to heart and have since upgraded your equipment. That is a fixed cost, hardware. The soft cost is your staff's productivity; let's talk about what you can do to boost it.

How efficient are they on the computer? How much time have you invested in developing their skills? Have you ever given the basics courses? Made sure they are very comfortable and confident in the things they are doing dozens and dozens of times per day? Don't assume that your staff is as comfortable on the computer as you are. All firms have the range of skill sets, you've got some people that a very tech-savvy, others at the other extreme end of the spectrum and

www.GlobalMacIT.com/boost   440-252-4600

everyone else fall somewhere in between. Even the advanced users have told us they got a lot out of our training. One small distinction or new tip can save someone minutes per day. When they these things 10X per day, it adds up significantly.

Email, calendar, contacts, PDF management, Word, saving and finding files, case management. These are all tools that people access on a daily basis. If they have figured it out on their own, I can guarantee you that the majority of your staff is currently wasting hundreds of hours using inefficient ways to complete tasks on their computer. Wasting 10 minutes a day, works out to 40 hours per person, per year. Think about that. 15 minutes a day is 60 hours per year. Small office with 5 people earning an average of $50k. That works out to $7,500, add on the fully burdened cost (taxes, etc.) that comes out to right around $10k in added profits added to your bottom line.

"Big deal. They know computers, they're doing fine. My staff is so swamped, I can't justify giving them 'computer training.'" That is EXACTLY why you must give them training! You see most people's computer training goes like this:

1) Congratulations, you have the job.
2) Here's your computer.
3) Get to work.

How often will a new hire confess his or her computer inapt ability right when they've just been hired. Never. So it is your responsibility to train them on an ongoing basis.

So ask yourself - when was the last time you blocked off time to help your firm develop their skills in the tasks they repeat multiple times, every day? We provide a monthly "Mac2Basics for Law" webinar for all of our clients and their staff. Our largest firm, with 37 users, has rescheduled their lunchtime on the day of our webinar, buys lunch for their staff, and they all watch it together because they have seen firsthand the results and positive feedback they get from their staff attending these trainings. Make this a priority for your staff and you will directly increase your bottom line. In addition to learning better ways to use their computers, by training and investing in them, you will make the more comfortable and confident in using their computers, which will help them in everything they do.

CHAPTER SIX

PROCESS MAKES PERFECT PART I: EXTERNAL

As I continually remind owners of law firms, you're very fortunate to be in the fourth most profitable industry in the nation. However that doesn't come cheap, as you also have the third highest payroll costs in regards to the overall revenue. This is a double-edged sword because it allows you to run very a poorly managed law firm and still make a good profit (and there are hundreds of examples out there). The sad truth is that you, reading this right now, most likely know plenty of attorneys running firms like this.

Today I'd like to discuss a very big issue I see happening over and over again. The problem here is hiring people. In times of growth or challenge, most firms simply throw people at a problem. I'd like to challenge the approach of hiring people and offer an alternative. Instead of throwing people at a problem, focus on the process.

Question: What do most firms do when they start growing?
Answer: Throw people at the challenge, right?

I want to challenge that approach. When growing, a firm's first instinct is to place an ad and then add a body. The thinking is that "another person will allow us to give them X,Y, and/or Z to do and free up someone's time". However, this instinct can severely impact the profitability of most firms because when a clearly-defined process is lacking, it is impossible to get your maximum ROI out of an additional body in the office.

There will be two parts to this article: the 1st part will focus on leveraging external processes while the 2nd will focus on internal processes. Both articles will provide you with direction on increasing your profitability, so get your highlighter out.

The first alternate option to hiring people when growing is to outsource. The primary benefit of outsourcing is that you get to leverage the focused expertise and developed processes that another company has developed. Let me share a couple examples of how we have done this.

RUBY RECEPTIONISTS

In 2014, we became very aware of an issue that we learned most people detested and which we, therefore, wanted to get rid of the phone auto-attendant. The alluring benefit to this is to reduce distraction from people having to answer every call coming in and to allow people to get directly to the person or department they need to reach. The reality of it, however, is that everyone despises these automated systems. Have you ever experienced the hair-pulling experience of simply wanting to talk to a human and hearing this: "Sorry, 0 is not an option, please try again."

Furthermore, when further analyzed, using an auto-attendant may have been costing our business thousands of dollars in lost business. Have you ever given up when trying to call a business and then called the next one down the list because you could not reach a human? So, we came to the decision that we needed to answer our calls live.

OPTION 1 – USE AN EXISTING BODY

Another option was that we could dedicate one person in our office to answer all calls. Every one of our employees more than enough work to do, so we quickly nixed that idea.

OPTION 2 – ADD A LOCAL BODY

The 2nd option was to hire a secretary to answer calls. Bringing on an additional employee would bear the full cost of wages, plus the fully burdened cost and overhead of having an additional employee. These costs add up quickly, with upfront hardware needed such as a computer, desk, chair, phone, taxes, increased insurance costs, etc. Then you've got the added monthly expenses for all the accounts and services they need (phone, email account, benefits, etc.) You also have to train someone and define the answering process. In addition, we did not have enough calls to keep someone on full-time, so this was really not a good option for us.

OPTION 3 – ADD A REMOTE BODY

We then looked to outsource. We did our research and then dug into the companies we were considering, and our research showed Ruby Receptionist to be the best option. They had been around for a while and had established processes that almost ensured quality. We now spend $250 a month with them and never think twice about it, as we get great feedback from everyone who calls in and all of our calls during business hours are answered by a live, human. We love it, and our clients and prospective clients love the increased customer experience.

UPSOURCED ACCOUNTING

Another example occurred at the beginning of this year when we looked at the cost of our traditional CPA firm we used to do our taxes and then added up the total time that our Office Manager was spending doing the books, plus our cost of payroll. We were not thrilled with our CPA firm, and they really did not add much value or take proactive measures to help us. We determined that our Office Manager, Chris, was spending 20-30 hours a month dealing with tasks related to bookkeeping and taxes. Most importantly, just because Chris could technically do the books, it did not mean that it he should. This was not his specialty after all, and he really did not enjoy doing it at all. We are too small to warrant a full-time bookkeeper here, so we started looking at outsourcing options.

We ended up finding a company called Team Incline, who is a flat-fee company that takes care of your bookkeeping, payroll and taxes, all for a flat, monthly fee. Our cost is $550 a month, but this now includes both the business and my personal taxes (previously paying $175/month), as well as payroll services and all of the bookkeeping. Chris now has freed up 20-30 hours a month (at $18/hour) to invest in marketing. Chris loves marketing and is great at it, and this generates a real ROI for our company.

On top of all that, the biggest benefit we feel we are getting is leveraging their expertise and developed, streamlined processes. When they took these responsibilities over for us, it felt SOOOO good, knowing that it was now being done better than ever before. It was literally a weight off our shoulders, freeing us up from focusing on this and allowing us to do what we do best. They come to us proactively, with things we would never have the foresight to think about. By leveraging their developed processes, we are benefiting in ways we never could if we had someone continue doing this task in-house.

These are just two examples in which we at GlobalMac IT have been able to outsource work to an external vendor. Our biggest benefits are freeing up time for our internal staff so that they can focus on higher ROI generating activities, and keeping our costs down from having to hire additional staff while leveraging the developed processes that they have developed since we are hiring them to do what they do best.

Side-note: It just dawned on me that the term outsourcing, for many, will make them think of outsourcing jobs overseas. The "times they are a changin'" and while there are many ways to do that, there are tons of US-based businesses who have evolved their business models to be able to work with such a model. Ruby Receptionist is based out of Oregon and Team Incline is based out of Ohio. I wanted to point this out to make sure readers don't close off this idea due to a negative association with the term 'outsourcing.'

WHAT WE DO FOR MAC-BASED LAW FIRMS:

We take on the entire scope of technology management for our clients, from small day to day issue resolution to full scale project work affecting the big picture for our firms. They outsource password management, new employee account setup and training, continuing technology training & education, hardware upgrades & secure retirement of computers, becoming the outsourced IT department for each firm. Our clients leverage our passion for Apple computers, our core purpose of redefining the relationship between people and technology, and our commitment to our core values (Helping Others, Constant And Never Ending Improvement, Sharpening the Saw, and Work/Life Balance) in every interaction we have! We have repeatable, detailed, play by play processes which we follow for all of our clients and when they hire us, they instantly get to tap into the years of experience we have supporting just Mac-based law firms.

WHAT IS IT YOU BUY WHEN YOU OUTSOURCE?

When you make the decision to outsource, you're not just buying the "service plan". Instead, you're making a decision to pay a price which includes the values and processes that company has developed over time. Consider those pieces carefully when looking to outsource any part of your business.

The big lesson here is to question where time goes in your firm. What is the actual cost to the firm for completing a task? Do you have a person spending 20 hours a month doing bookkeeping? Run the numbers, what is that overall cost to your firm? And don't just multiply their hourly rate, factor in their fully burdened rate, which is another 30% on average. However, look deeper. In addition to that number, what is the impact on that employee's productivity? Is this a task they enjoy or loathe doing? If that time was freed up, what is the overall impact of the other work they could complete? Finally, tack on the total upfront and ongoing costs of having an additional employee.

So, put down your highlighter and take out a pen, legal pad, and calculator. Re-read the paragraph above and work through the numbers. I'd love to hear about your results and any AHA! Moments you get from this. If you've made it this far and actually taken action (after all, this entire article is worthless if you're not going to do anything about), please email me at tom@globalmacit.com and let me know about your findings and you're going to do about it!

www.GlobalMacIT.com/boost   440-252-4600

CHAPTER SEVEN

PROCESS MAKES PERFECT PART II: INTERNAL

In the previous chapter, I discussed the benefits to leveraging outside sources that have well-defined and established processes, and taking this approach in certain situations as opposed to just hiring someone. The primary focus was outside processes. In this second part, we will dive into why 'process makes perfect' and how to go about creating processes in your firm. [Quick apology – this article ended up being longer than planned, but it is all very useful, so I am splitting part II into two parts, like splitting atoms. Like that? ☺] Alright, here we go!

Let's start off by making a case (no pun intended) for using processes in your firm. I remind the attorneys and firms we support on a regular basis that according to Financial Information company Sageworks, in 2014, Legal Services was the 2nd most profitable industry, with a 17.8% Net Profit Margin! Not bad. While that is great, I have found it creates a big problem; many firms can get away running a sloppy law firm and still make money. Another point to take into consideration is that the legal industry has the 4th highest payroll costs, coming in at 45.03% of revenues. Because this is such a huge cost, anything you can do in your practice to increase output while keeping your payroll costs fixed, will increase your bottom line profits.

This is where processes come in and will pay off big-time. Many attorneys bill for different work at different rates, so if you reduce the amount of time you spend on low-revenue generating work and instead delegate that someone else, freeing you up to focus on work which you bill at your highest rate. And although you have done something for years and can do it in your sleep, it does not mean that is something you should be spending your time doing. The absolute equalizer is time, so the more things you can delegate and remove off your plate, the better.

This is just as important for your staff. Do they have specific, well-defined processes that drive the things they do over and over or do they 'wing it' and do things a different way every time? What happens when an employee leaves or is let go? If you do not have defined processes, it will often take them a LONG time to get good at doing the things they need to do, and someone else will need to hop in and help.

Look at the flip-side: you have well-defined, play-by-play instructions for many things that are done, so you simply educate your new employee on where to find the instructions, then set them free and let them come to you when they need clarification. Wow, a world of a difference.

FIGHT THE URGE AND GET IT OUT OF YOUR HEAD.

When we started creating processes, it took a while to get started. Let me share the lessons I learned in the hopes that I can provide you with shortcuts to process creation. The most common issue is 'the urge': "I'm so busy I don't have time to create a process." 'It's just easier/faster if I do it myself." The urge is the belief that it takes too long to create a process and that you don't have the time to do it. In order to overcome this, you need to take a high-level view. Add up the total amount of time creating this process this will save you over the next 3, 6, 9, 12 months. Being nearsighted only takes the now into consideration, and your mind will think that it's easier to just do it yourself.

Whether we are conscious of it or not, we have a process for tasks we do repeatedly. By becoming more conscious of the process, we can start to define it and look for holes in our process. [I want to point out that a process does not have to be something we delegate. I have many processes I use myself which keeps me moving through it as fast as possible and ensuring I don't leave out any details.] Legal work is extremely detail-oriented, and a goof-up can easily become a costly mistake.

When I first started doing this, I simply had a legal pad and started writing down the steps as I did it. If you are doing a short process, like how to file in a specific website, you can probably just make that list and all info you need. If you're creating a complex process, such as hiring a new employee, you will want to get it all out of your head. Then you will need to break it down into chunks, which you can then dive into one at a time. As my process for creating processes has evolved, I prefer starting with a complete brain-dump onto a whiteboard. I personally work best with another person to bounce ideas off and to help pull things out of my mind.

As an example, we are currently revamping our new employee process for our clients. When one of our clients hires someone new, there are a tremendous amount of things that need to be done, from an HR perspective, IT, accounts, training and more. We have broken down how we are going to build this process in a series of phases:

1. Pre/Pre (makes sense to me) - This is a proactive call we will have with our client way before they hire someone, so that we can define what their HR process is and to see how we can best fit into this process to alleviate work on their plate and clearly define the things we will take care of and what they will take care of.
2. Form & Setup Project - Intake all the information generated in the custom form we have for our clients when they have a new employee. By making this detailed and thinking of all angles, we can gather the majority of the information needed up front. We use Typeform for this, an amazing tool that kicks SurveyMonkey's butt and makes it look and feel like a dinosaur! We then set up the project, create tasks and process all the information.

3. Purchasing & Licensing - if needed, we reach out to the Apple Business team to get a quote for the requested hardware, plus anything else they will need (additional monitor, battery backup to protect desktop, USB to Ethernet adapter for laptops, etc.) - again we have thought of everything: the outcome is zero reactionary actions once the new employee starts. We purchase all licensing and create all necessary account (Microsoft Office 365, Box for file access, Rackspace email account, etc.)
4. Tech Setup - Our technicians then install all software, register licenses, sets up the email accounts, file sharing access, VPN access, applies our custom Security Profile, and more.
5. Quality Control - 24 hours prior to a new employee first day, everything is ready to go, but we make no assumptions. We have a quick call with our client to review the system and make sure we did not overlook anything.
6. Training - We have created training 'drip' with a sequence of emails that will go out to the new user over their first month with tips and videos we have created specifically for the staff of a Mac-based law firm. Most employee computer training goes like this: "you're hired, here's your computer, get to work." We know that additional training will not only make the new user feel more comfortable and confident using the computer, but also more proficient, maximizing your work output from them.

As detailed as this may seem, this truly is just a quick overview. Each of the bullets above has a very detailed, play by play process and checklist created. Going through this process has allowed me to switch from providing a fair amount of tech support and doing almost all of the onboarding of new clients, to being able to delegate 100% of these responsibilities to my staff, all while keeping the same quality. I can now invest more of my time in activates that have a higher return on my investment.

WHAT PROCESSES?

One of my favorite business gurus is Verne Harnish - his book Scaling Up 2.0 is an absolute must and is most definitely applicable to law firms - at least for those who are running it as a business, as opposed to just practicing law. In one section, he has us define the key 4-9 processes driving our business, then assign someone to be in charge of tracking these. Once you have defined these, pick the one that needs the most improvement, or the one you need to get off of your desk the most. Then block off time to start working through and defining these. As an example, ours were:

- People Development (Recruiting and internal skill development)
- Sales & Lead Generation
- Onboarding of new clients
- Customer Satisfaction
- Best Practices (the sum of all the key areas we look at for our clients and the processes we go about implementing these)

- Service/Product - quality of our support, response time, etc.

You should plan to spend about 90 minutes with your staff to determine what these key processes are. For more direction in Verne's Methodology, pick up his book and flip to page 52. I am an avid reader and this in my top 3 business books of all time.

An alternative approach, as opposed to getting the 'big rocks' as I describe above, is to keep a diary log in 15-minute increments of what you do. After 1-2 weeks, grab a highlighter and pick the things that you could delegate or spend a considerable amount of time on. Both of these are good places to generate some small wins for you. The things you can delegate can quite easily be the basis for your employment ad. If you create processes for these things, you can then get them off your desk.

We'll wrap up here as you need to just do what I said above...get started! Come back next month and we'll talk about demystifying the process of creating processes and then how to test and refine the processes. You will not want to miss those steps.

DEMYSTIFY THE PROCESS > FORGET THE TOOL > START

Most of us will spend a tremendous amount time thinking about something we should do versus taking action on it. When I wrote my first book last year, I thought about it for months and months without ever writing the first chapter.

Once I started writing, I kicked myself as much since it was actually LESS draining to actually write and take action that to think about it. My concept of "Demystify the Process," which I say here to my staff (including myself) all the time, is a reminder that thinking about something does NOTHING. You need to simply break it down, stop thinking about it and just take action.

Ok, you're ready to start and are now sold on this idea and realize you need to start creating some processes. If you're anything like I was at this point several years ago, you may be tempted to wait starting until you've found the PERFECT tool to help you track and create this process. I literally spent a couple months googling and searching for the Holy Grail of tools to help me create processes. I finally got myself unstuck, said screw it and opened a blank Word document and started writing the steps, one by one. My team started doing this, and I made it a requirement to create two processes a week, big or small. Compound that and we now have hundreds of processes, which ARE helpful, and we use all the time. Don't get stuck on finding the right tool, focus on starting, you can always find a better tool later. To save you some time, now we are several years into this process creation, our two main tools we use are Confluence by Attlassian, Manifest.ly and Affinity Live. Bottom line, however, is to not delay in getting started in creating processes because you are waiting to find the best tool.

Finally - START. Get yourself unstuck and take some kind of action. Start small, and work your way up. The more you do this, the more natural it will become. Get your team bought into this and your productivity will soar.

TEST & REFINE

The last part of all this is to test & refine your process. Although it may make sense in your head, the ultimate test is to give it to someone else and ask them to do it. This is where you will find out where the holes are. To the chagrin of my wife, we lost our best-ever nanny a couple months ago as she graduated and moved away. The new gal, put our kids to bed after lunchtime the first time just a few days ago. My wife, being very detailed driven, provided her with a one-page summary of everything that she needed to do. One of those things was to bring our 2-year old son to bed at 1pm and read him 3-4 books. Well, she did this. However, he didn't go to bed till 2pm; she read him every page of a 60-page book. So instead of reading to him for 10 minutes, she read to him for an hour. It's not her fault, she didn't know, she followed my wife's instructions to a T. My wife has since clarified this step in the process.

Having someone else execute your process will find the kinks in it, and you will then further clarify it until you can fully hand it off. For detailed processes, I prefer to be looking over their shoulder their first time through to do this in person, so I can tighten up my process. We provide all our the firms we support with a $1,500 cloud-managed security appliance since most smaller firms tend to be using a $100 Best Buy router, which is not meant for enterprise and does not provide the level of security needed. I used to spend 2 hours setting this up every time we have a new client since I was the only one that knew how to do it, and it was a tedious, detailed process. I forced myself to spend a little more time to create a process and after going over once and tightening it up, my entry level tech can now set these up, 100% on his own and we have had zero issues over the past few months as we have brought on multiple clients. This has freed me well over a dozen hours in the past 6 weeks alone, just on this one process.

Becoming a process-driven law firm will increase your profitability in a very significant way. In the past, we could handle about 50 desktops per technician. Thanks to everything we have done in this arena, we can now handle 75, and I have no doubt we can get to 100 per technician. This ends up adding a tremendous amount to our bottom line and profitability, not to mention having a much better atmosphere with very little reactive, cleanup work. Thanks for bearing with me through all process hoopla. If this was helpful, and you've bought into all this, I challenge you to grab a pen and write down one process you will create in the next week that you have been putting off, but wanting to do for a long time. Block off time in your calendar to do it and simply do it.

CHAPTER EIGHT

WHAT OPTIONS DOES A MAC-BASED LAW FIRM HAVE FOR IT SUPPORT?

One of the most common things I like to educate Mac-based law firms on is what option DO they have for IT support? Let's assume you realize you shouldn't keep wearing the IT hat and that you are ready to focus on running your law firm, where do you go?

My hopes here, are that, at a minimum, you'll walk away with some very useful insights, which could very possibly change the way that you deal with IT in your practice. I am confident that this section alone will prove to be valuable and will arm you with helpful information.

So, that being said, every law firm pretty much has 3 options, although I have found most firms really only use 2 of them. The first, I covered in detail in the first chapter: Self-provided IT – what I call "wearing the IT hat." The second option is using an hourly person and last is Managed Services. Let's dive into each once.

SELF-PROVIDED IT – 'WEARING THE IT HAT.'

Many of the smaller prospective law firms that come to us have been taking care of IT on their own. As I had elaborated on, while it may initially save them money, the truth is, in the long term as the firm grows, it actually ends up costing far more than if they were to outsource their IT support. If you still think it's a good idea for you to handle your firm's IT needs on your own, please go back and re-read chapter 1 ;)

HOURLY SUPPORT

Most people "wearing the IT hat" will have an hourly support person they call on when needed. Some use them for IT related tasks, but what I have found to be most common is that they will call this hourly person when they are out of ideas, in dire times.

I started doing hourly support back in 2007 and did this for several years. The biggest limitation of this is that when you call someone out to fix issue A, they come out and fix that, to prove they completed what you asked of them. BUT even though they might come cross a laundry list of other items, it is not in their best interest to go ahead and fix all the issues they come across, which would then put them over the hours they had initially quoted you, putting them in a bad light. For example, an attorney would call me and explain their issue. I'd say that should take an hour, but when I start looking into it, I see much more that needs to be done. I couldn't just go and work 4 hours to fix everything.

In addition, since they only get paid by the hour, the hourly consultants need to do their best to line up their days with appointments. Their focus is to complete what you asked them to do, then need to run to the next job they had scheduled. I remember being in these shoes and constantly having a nagging, subconscious voice tugging at me, "In a perfect world, I wish I could keep an eye on all their systems and be more proactive, since it would overcome so many issues before they even came up." The truth is, the hourly consulting approach does not allow that to ever happen, and, in most cases, it ends up being a glorified reactive support model, where people only call in dire times.

With this being said, the hourly support model does not:
- Look for proactive, long term fixes on your IT network
- Allow time to develop Best Practices for your practice, since the tech is always just focused on completing the task at hand.
- The hourly tech does not get rewarded to establish maximum uptime and to increase the overall productivity of every employee in the firm.
- More often than not they work with anyone with a computer, hence have zero specialization in regards to law firms.
- More often than not, they are sole proprietors, often part-time workers, which means they often not available when you need them the most.

MANAGED SERVICES

The next option when it comes to IT support, is Managed Services. This is the opposite of the reactive, hourly model. The biggest benefit is that it helps to budget IT costs and shifts the support from a reactive to proactive approach.

The majority of Managed Service offerings you will come across are offered in a 3-tier model, such as Bronze, Silver and Gold. Bronze, for example, would only provide remote phone support, for example, with Silver providing more, etc.

Problems

Some of the problems with most Managed Service offerings are related to the tiered system. This means some things are included and others are not. Items such as Project work, training, after hours or emergency support, new user setups, office moves, on-site support, etc., are often additional.

This can lead to very confusing billing and making you feel like you are being nickel-and-dimed even though you are paying a monthly fee. It ends up being a nightmare for many clients, trying to track what is billable and not, and ends up acting as a hindrance for the firm staff to get support.

This creates uncertainty whether something will be an additional cost or not. If an employee thinks he's going to generate an additional hourly cost, they often don't voice it and just deal with it. This is similar to looking at health insurance options; where it is very confusing to grasp what you will end up paying; no matter what you choose, most people feel like it's a crapshoot; your fear both overbuying or under buying and have little confidence in your decision.

You see, when a new client is taken on, it is not uncommon that it takes an average of 5 hours/computer in terms of up-front project work (usually clean-up work to stabilize the client). This means there is a large additional up front cost to get started. So in addition to your monthly Managed Services fee of $1,500, for example, you may have to pay $3-4,000 up front.

In the future, anytime project work is needed, whether it is an office move, installing new software or hardware, setting up a new computer, etc., it is usually billable. This is rarely budgeted ahead of time, so it ends being an additional cost on top of the existing IT budget, throwing off your numbers.

I don't know about you, but I like to have all expenses budgeted out and on a monthly fee. No one likes big, unexpected expenses added to the budget.

For example, let's say you wanted to implement a new case management software, which would take 30 hours for planning, implementing and training. This would definitely be a billable project for most Managed Service plans so you would need to budget for that, at $150 an hour, $4,500 in labor, plus the software cost. At this point, they would either have to budget this out and delay the implementation. Or, the firm may decide try to do it themselves, theoretically saving the project cost.

The problem here, which we have often seen, is that when a firm tries to roll out a new solution on their own, they are not able to dedicate the proper time to research the solution, prepare the installation, then also plan for the training. Many times, they spend a lot of money on a solution, but they end up getting very little buy-in and usage. The main reason for this is that they do not have experience in rolling out a new solution, making sure it is setup properly from the get-go. Another commonly overlooked and critical part is having the proper training is performed, on a an ongoing basis, making sure all users and develop confidence and comfort in using this solution. We see very low buy-in every time one of these steps is skipped, so you end up with a very expensive Rolodex that people hardly use.

The fact is that many projects like this, which would increase productivity or increase your uptime and stability, end up being delayed because they need to be budgeted. This hinders the growth and advancement of your firm and hence negatively impacts its bottom line.

WINDOWS-BASED MANAGED SERVICES COMPANY

The truth about Managed Services in the Apple World is that it is still fairly new and there are less than 100 all-Mac Managed Service Providers in the nation. This means that many Mac-based law firms looking for a proactive support option often only come across Windows-based Managed Service Providers that SAY they can support Macs. Now, I don't blame them, because business is tough and most cannot justify turning away the business (although, in my opinion, they ethically, should say "No, we cannot provide the same level of support to Macs as we do Window.")

Problems:

The number ONE issue with these companies providing Mac support goes right back to the hobbyist principle we discussed in "Wearing the IT Hat." Overwhelmingly, it is financially impossible for a Windows-shop to provide the same level of support to Macs as they do to PCs. They simply cannot justify spending the same amount of resources on training, tools and support for the 5-10% of their computers that are Macs. Because of this you will never be able to receive the same caliber of support they provide to their PC users.

Sadly, we've seen many cases where their "Mac Certified" person is not Mac Certified at all. For example, one person had only sold iPhones over a holiday break at Apple. Another, the CEO of an IT Managed Service Provider that one of our clients had been speaking with had taken a self-paced, free online test on how to connect a Mac into a Windows network, then claimed he was Apple Certified.

As much as I hate seeing, I have to be honest that we we run into a Mac-based law firm who has been previously taken care of by one of these Windows Managed Service Providers who said they can do Macs, we knock it out of the park. Also every time we hear that the client feels they know more about Macs than the IT company. Or the IT company has one "Mac"

person and if that person is busy or out, they are out of luck and no one is able to help them.
SUMMARY OF IT SUPPORT OPTIONS:

All of these IT support options share one common trait – they are all generalists in their business:

I. If you are taking care of IT yourself - you're an attorney first and an IT hobbyist at best.

II. The main driver of the hourly support person is the billable hour, so they are never able to learn about all the needs of any specific type of industry, since they typically support anyone with a computer.

III. The majority of Mac-based Managed Service Providers support any businesses based on Macs. My company, GlobalMac IT, is the only company is the world that only supports Mac-based law firms and nothing else. Legal is not one of our verticals, it is the only thing we do.

IV. And lastly, Windows Managed Service Providers are all Windows first and Mac hobbyists at best.

CHAPTER NINE

WHAT ARE MAC-BASED LAW FIRMS SAYING ABOUT GLOBALMAC IT?

BEST IT DOLLARS I HAVE EVER SPENT!

As a new Mac user it was with great pleasure that I discovered Tom and GlobalMac IT. Over the years I have encountered numerous IT professionals but none came close to GlobalMac IT. Their efficient, prompt and highly proficient skills diagnosed and remedied several issues in a short period of time. Furthermore, Tom analyzed how I use my Mac and then made recommendations to improve my productivity. The best IT dollars I have ever spent.

Mark T. Bradshaw, San Diego, CA

Ben Stevens
The Mac Lawyer - The Stevens Firm

I HAVE TO LET MY FELLOW FAMILY LAW ATTORNEYS KNOW OF A HUGE MISTAKE I MADE FOR THE FIRST TWO DECADES OF MY LEGAL CAREER – HANDLING MOST OF MY FIRM'S IT NEEDS MYSELF.

For instance, when one of our computers had a problem (and when we used PCs there were many), I thought the best course of action was to try to solve it myself instead of having a professional do so – as it would "save money." All of that changed late last year when we (finally) realized that the best course of action was to outsource certain functions in order to allow us to spend more time doing what we do best – handle family law cases. We started using GlobalMac IT for our computer and network needs, and we are beyond pleased with the results. They understand how to setup computer networks for law firms, and I don't have to worry about issues arising, if they do, I know the team at GlobalMac IT will be there to save the day.

THE "RETURN" ON THE INVESTMENT WITH THEM IS THAT YOU CAN SPEND YOUR TIME PRACTICING LAW INSTEAD OF WORKING ON IT ISSUES.

The other big plus from using them that is easily overlooked is the preventative maintenance that they do. They monitor everything and keep your software up to speed before problems arise, which prevents downtime, lost productivity, lost income, etc.

I had a tech guy here in Spartanburg that I used to use, and he was good -- when I could get him and only after problems arose. I wish I'd gone with GlobalMac IT a long time ago, because I don't worry now about our IT needs. For instance, we decided last week to go with DocMoto for our document management needs, and all I had to do was call GlobalMac IT and have them get in touch w/ the DocMoto guys, and they take care of all the tech stuff behind the scenes -- which saves me hours upon hours of time messing with it.

So the real question is, **do you want to hire GlobalMac IT and let them manage/maintain your system for you so you can practice law, or do you want to wing it** (whether doing it yourself or hiring someone local to deal with problems when the arise. I think that using GlobalMac IT is (by far) the best option for professionals and that you'll agree a few months down the road after you get over the initial costs. Buying a top notch router, computers, etc. is all expensive (I've bought 3 iMacs in the last 3 weeks, so I know), but telling GlobalMac IT, "hey, we have new employees and new computers" and then having them configure and set everything up behind the scenes is great.

Ben Stevens The Mac Lawyer
Attorney
The Stevens Firm, P.A.
Spartanburg, SC

www.GlobalMacIT.com/boost   440-252-4600

Colvin, Chaney, Saenz & Rodriguez

"THEY KNOW HOW TO
TAKE CARE OF OUR EVERY NEED.
MY STAFF LOVES WORKING WITH GLOBALMAC IT."

I have worked with every team member at some point but on a daily basis I speak to Thomas and Jeff and they are great guys. Not only are they friendly and always in a good mood, they know how to take care of our every need. My staff has nothing but good things to say about them. Congratulations on having such a great team!"

Mary Martinez
Law Office Administrator
Colvin, Chaney, Saenz & Rodriguez
Brownsville & Edinburg, TX

"OUR PREVIOUS IT GUY NEVER TOOK
THE TIME TO FIND A SOLUTION."

"Everyone is very patient and does whatever is necessary to fix any problems we have. I've been dealing with many problems for a long time and our previous IT person never took the time to find a solution. I am very appreciative."

Lety Allen
Paralegal
Colvin, Chaney, Saenz & Rodriguez
Brownsville & Edinburg, TX

Desert Law Group

"GLOBALMAC IT OFFERS SOLUTIONS WHICH IMPROVE OUR PROCESSES AND PROCEDURES FOR OUR LAW PRACTICE, MAKING OUR ENTIRE STAFF MORE PRODUCTIVE, AS OPPOSED TO JUST FIXING THE INITIAL PROBLEM."

I can't say enough about the prompt response GlobalMac IT continues to provide us in regards to our IT related issues. Their business model allows them to take a proactive approach, suggesting solutions, which improve our processes and procedures for our law practice, making our entire staff more productive, as opposed to just fixing the initial problem.

When we first started transitioning to their services, several of our team members had dealt with repeating issues, which our previous IT provider was unable to resolve. The GlobalMac IT team was patient, thorough, and successfully resolved all of the major problems we had been dealing with for a long time. I also did not have to worry about being charged for each individual support call because we have a truly all-inclusive support plan that encompasses anything IT related!

Kimberly Lee
Attorney
Desert Law Group
Indian Wells, CA

www.GlobalMacIT.com/boost   440-252-4600

"WHY YOU MUST SWITCH YOUR LAW FIRM TO MACS AND WHY GLOBALMAC IT IS THE SUPPORT PARTNER YOU NEED"

For eight years we struggled with PCs and PC service providers. The equipment kept costing more and more and the service providers kept going out of business. Every problem was a "crisis" that required hours, if not days, to fix, IF we could even reach the service provider at all. Quite often they would say they would be available in about 72 hours, as if it was okay for us to be down that long. Proactivity was a pipe dream and I was spending a lot my own time, every day, trying to fix issues, instead of running my law firm.

The Apple Business Team in Cupertino connected us with GlobalMac IT when I was ready to migrate over. Within 30 days everyone was acclimated to the change, and things were much simpler. Tasks that took a lot of time on a PC took substantially less on the Mac. Large problems are now almost unheard of and when we do have issues or need support, GlobalMac IT is always there.

Many times, support tickets are fixed the same day. No longer do we feel like we are imposing on their time like our previous IT providers; instead they continue to tell us that they are here to help us and make us more productive.

Looking back 3 years later, we cannot imagine going back to PCs or leaving GlobalMac IT. It has been one of the best decisions we made as a company.

Stuart Horwitz
Principal
Horwitz & Damincone
Akron, OH

Rose Joneson Vargas

"THE INCREASED EFFICIENCIES THAT GLOBALMAC IT HAS IMPLEMENTED FOR OUR TEAM HAVE MADE A DRAMATIC IMPROVEMENT IN HOW OUR FIRM OPERATES."

The implementation of file management solution has been a huge benefit for our firm and our clients, specifically given the bandwidth limitations we have on the island. The work directory system you devised and implemented for us is well organized and much more intuitive to use than our previously setup. The ability to access our files remotely is fantastic and of course the constant back up and syncing that GlobalMac IT conducts and monitors provides us peace of mind. Lastly, the ability to share documents with our clients and other parties has been a great tool for us and one feature that we are utilizing more and more each day.

We are currently working on a very large and complicated case involving attorneys in Seattle, New York and Madrid in which we have had to share a large number of documents, some of which are quite large. The ability to set up a secured shared folder which automatically syncs to our work directories has let us get these documents to the attorneys quickly and in a very organized manner.

Not only is it efficient but it puts our little island firm in a very good light with the large multi-national law firms we are dealing with and our clients who are scattered across the U.S. It is hard to imagine how we would have taken care of all the different aspects of this case, the re-organization you did to

www.GlobalMacIT.com/boost   440-252-4600

our Work Directories and the implementation of the file management solution.

The increased efficiencies that GlobalMac IT has implemented for our team have made a dramatic improvement in how our firm operates.

Barry Rose
Managing Partner
Rose Joneson Vargas
Pago Pago, American Samoa

Hajek & Beauclaire Attorneys

"BY CONVERTING MY LAW FIRM FROM WINDOWS TO MACS, WHAT GLOBALMAC IT HAS DONE FOR MY ABILITY TO LIVE MY LIFE AND LAW PRACTICE, HAS BEEN REMARKABLE."

In January 2010, I decided to convert my law offices from PC's to Apple and to become a "paperless" firm. My firm has offices in several states, and I met Tom while working in my California office. Even though there are plenty of Apple consultants in Minneapolis, I was so impressed with Tom's skills and abilities that I flew him to my Minneapolis offices for a week to do the conversion with my entire staff.

In the following months, Tom worked efficiently with my staff to train them and correct any problems that were encountered. People acclimated to PC's are often resistant to converting to Mac, but Tom's team worked with my staff for what seemed like an effortless and seamless transition.

GlobalMac IT has been very responsive to any questions or concerns any of us have had. The bottom line is that

GlobalMac IT is the most competent IT services firm I have ever worked with, and I highly recommend his services to all Mac-based law firms.

By converting my law firm from Windows to Macs, what GlobalMac IT has done for my ability to live my life and law practice, has been remarkable. For instance, I am in Paris and my secretary is up in her cabin in the middle of nowhere. She is tethered to an iPhone, and my partner is home with a sick kid in Minneapolis. Yet, we are doing business at a high level; it is instantaneous.

The setup and tools have to work to remain productive. This does a few things for me:

1. It provides good employees; because you have to trust the people you are dealing with.

2. You need to be able to monitor what they are doing, and the tools you have implemented for us lets us see in real time what is being accomplished.

3. The employees appreciate that they do not have to be tethered to the office if they get sick, or if it is raining or snowing. The work goes on, which is critical.

That is what the services of GlobalMac IT have allowed my business to do, without that I would be back in the Stone Age. I literally could not function the way I do without what you have done for our law firm.

Bob Hajek
Attorney
Hajek & Beauclaire
Minnetonka, MN
Del Mar, CA

www.GlobalMacIT.com/boost   440-252-4600

INCREDIBLE FREE SPECIAL OFFER…

Step #1:

Complete the IT Survey Questionnaire at: www.GlobalMacIT.com/itsurvey

Step #2:

After completing the IT Survey, you'll be sent a link to my schedule.

Step #3:

Select a timeslot for your **FREE, 60 MINUTE, Personalized** Consultation to discuss how to boost your firm's profits through an IT Transformation.

Valued at $497.00!!!